Superior Humanity
New World Order

Table of Contents

- 0: Introduction 2-3
- 1: What is life? What is death? 4-7
- 2: Religion 8-15
- 3: Science 16-22
- 4: Aerodynamics 23-27
- 5: Electrodynamics 28-33
- 6: Metadynamics 34-38
- 7: Timespace dynamics 39-44
- 8: Thermodynamics 45-49
- 9: What is the fate of humanity? 50-51

For the love of my family

(Image from Shutterstock)

chapter: 0

INTRODUCTION

"Zero is light and it has no value of mass." Chakotay B. Johnson

Hello, this is a non-religious approach to bring ancient alien humans together in peace, equality, and liberty. Mother earths, father trees, and children seeds. I am theoretical and dynamical physicist Chakotay, "he who walks the earth facing the sun, sky, and seas." If humans come together in peace, they can achieve immortality. I want to let the readers know that this book is not about me, instead it is about the entire humanity. Humans together as one is "god who is king." I want to thank Albert Einstein, Nikola Tesla, Elon Musk, Andrew Powell, Martin Luther King, Malcolm X, John F. Kennedy, and Steven Hawking for their advancements in science to help the superior human species.

(Image from Wikipedia)

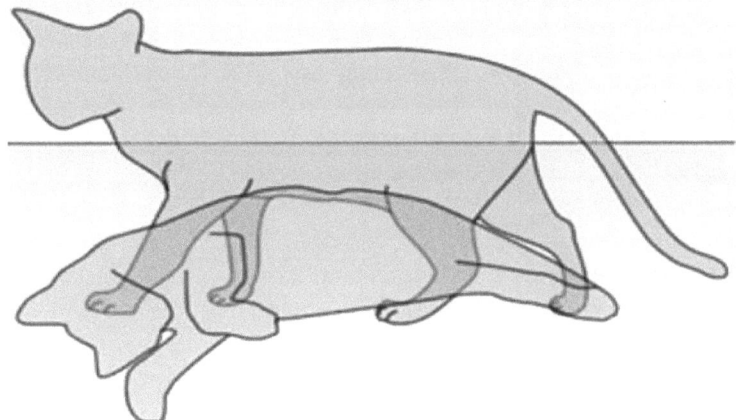

chapter: 1

What is life? what is death?

"One is unity and the greatest thing for humanity." Chakotay B. Johnson

Life is an animated thermodynamical biochemical organic process made up of approximately thirty trillion (3×10^{12}) cells, all having a unique function and purpose. Human beings are omniscient objects that materialize from darkness and pure nothingness of quantum consciousness. What modern scientists call unconsciousness is the initial state of all existence where consciousness and physical reality of superior humans materialize. Quantum consciousness does not experience time. In quantum mechanics photons, or particles of light, are in a simultaneous state of the past, present, and future being unified. In physical reality humans experience time.

(Image from google)

Superior humans are avatars of the observable universe, with a dualistically gentle and destructive essence to their existence. The universe is made up of seventy-three percent (73%) hydrogen and twenty-five percent (25%) helium, and less than two percent (2%) of heavier elements that form organic life forms. The rareness of these heavier elements and the fragile nature of living existence, humans must co-exist in peace and depart themselves from violence and self-destruction. The gentleness of human existence is a true blessing. All humans are equal and have a true purpose. The universe exists within us. Superior humans of planet earth are diamonds (heavier elements) in a heap of stones (hydrogen and helium). In a universe of infinite scenarios and possibilities, superior human potential can achieve anything. Einstein said, "Imagination is more important than knowledge. Knowledge is limited. Imagination encircles the world." Humans have limited scientific knowledge. On the other hand, they have infinite imagination for continuous scientific innovation. Only working together as one, humans can achieve quantum greatness. Reality is what humans want it to be. Us together we can have anything our hearts desire. We are the all-seeing!

(Image from Arstechnica)

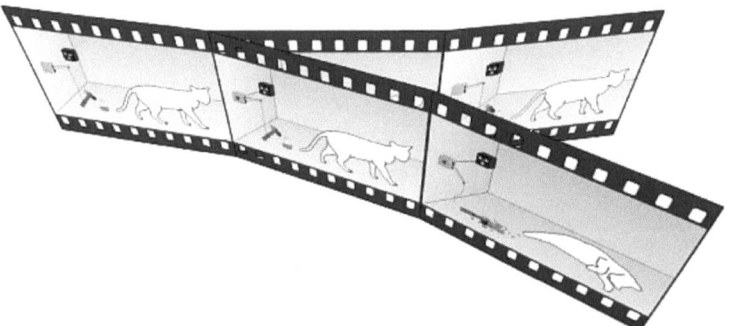

The observable universe is a multiverse, or multiple universes, where everything is predestined and predetermined based on decision and quantum consciousness. An unstoppable intelligent force of unified consciousness is the master of the day of judgement. Humans must rise into total peace or fall into destruction. All humans are a part of a greater primordial quantum consciousness that always existed. Humans must accept the diversity in other beliefs and opinions. Humans are not made the same because the complexity of thinking, and the surrounding environment. Being alive is only a product of our physical imagination in a multiverse of infinite existence and quantum reincarnation. Human existence is like a cyclic wave of energy. It can stay at its high (crest) of life, or its low (trough) of death, or it can go back and forth.

Death is interpreted as the end of human experience of life and the quantum equilibrium state of being conscious. It is not the end of their story. Humans are conscious quantum energy that has existed for all eternity. When humans die and lose awareness of reality, their consciousness return to the unified primordial quantum consciousness and all-encompassing energy of light. In the quantum state of being dead and unconscious, the desire to live again and infinite imagination causes the dead to rematerialize into some type of form of life. So, death is ultimately an illusion in the multiverse. May those who read this entire short book learn the true power of the superior humans of the new world order.

(Image from tribune)

chapter: 2

Religion

"Two is for the dualistic state of the human mind." Chakotay B. Johnson

(Image from Wikipedia)

Religion	Percent
Christianity	31.11%
Islam	24.9%
Unaffiliated	15.58%
Hinduism	15.16%
Buddhism	6.62%
Folk Religions	5.61%
Other Religions	0.79%

Size of Major Religious Groups, 2020[1]

Religion is a unique or similar social-cultural system of beliefs and worship of something or someone greater. It ascribes divinities, behaviors, practices, arts, texts, groups, and locations to a higher power or way of life to give humans a deeper meaning. There are many religions. For centuries, religion has turned humans against each other. The superior human mind is infinitely dualistic and conflicting. Having radical views about your particular religion fuels hatred or distaste of other religions that is extremely dangerous for the delicate existence of superior humans. Humans should respect each other's beliefs and opinions. No religion or culture is better than a superior human life. Superior humans working together has the divine power of intelligent design. Any technology created throughout the history of humankind will never be able to out-think the superior human mind. This is new world order(non-religious), we are superior thermonuclear ancient solar humans. Terrestrial sentient alien beings in a multitude of alien societies.

(Image from Britannica)

Christianity is the world's largest religion. It is a sacred religion that always existed in a trinity with Islam and Judaism. Jesus Christ, or Yashuah Elah is the only begotten son of God and who so ever believe in him as the way and the light shall be granted eternal life. Jesus was born here on earth, healed the sick, resurrected the dead, and fed the poor. Some say he died, then rose three days later. And some say he went to heaven and is watching down on humans until the day they are sent back to him to be judged. According to quantum physics this is pure fact. We were not put here to fight; we were put here to unify. Jesus lives inside the light, so humans building the quantum star gate using the first and second law of Thermodynamics, explained in chapter 8, and the third law of Timespace dynamics, explained in chapter 7, will resurrect him upon superior humankind. It is all within the mind.

(Image from cfi)

Islam is the second largest, and fastest growing religion in the world. It teaches unity, and that everything in existence is one unit of creation. It is solely against the exploitations, manipulations, oppressions, and deceptions amongst superior humankind. A Muslim's destiny is peace with all of his/her surroundings. A huge misinterpretation in Islam is the treatment of women; the Quran says treat them with love and equity. In the Quran Allah is the "Us," and the "We," who really is all superior humanity in unity and peace for all eternity. Also, in the Quran Jesus is an afro-asian light monk who was not sacrificed on the cross; It was made to appear he did, though. In the multiverse this is accepted as highly possible. He escaped from persecution into Asia where he spent the rest of his days in love, meditation, and peace. To Muslims, and certain Christians, and certain Jews, alike, Islam is not a religion, but instead a way of life.

(Image from Britannica)

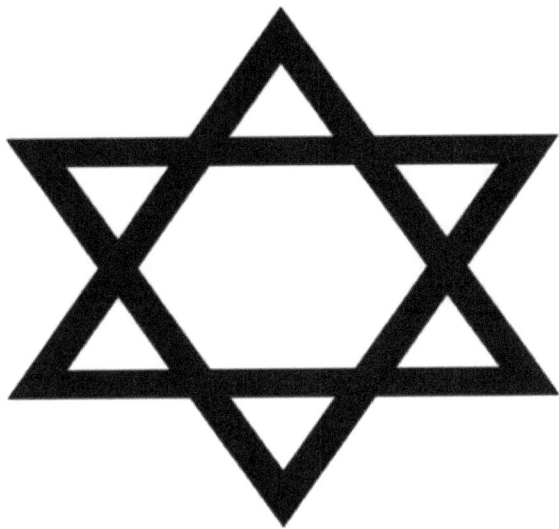

Judaism is the religion of the "Children of Israel." Although it is an exclusive and small religion, it is in the synchronicity of the holy trinity. The "Children of Israel" are the chosen people to save all humanity. They are the sons of God and the daughters of man. Truth be told, they are no different than any other group of superior humans. The chosen "Son of Israel" is Jesus Christ, his destiny is to save all of superior humankind. Jesus Christ was the king of his people, and he is an expression of the chosen people who are destined to rise into an eternal state where all superior humans are equal. Planet earth, where superior humans call home, must build a superior human utopia of liberty and justice for all.

(image from Mahdi foundation)

Quantum Mahdiism is a religious and scientific system of beliefs. It teaches that everything in our physical reality can be described with quantum information derived from the superior human body. It also teaches that every religion and people always existed as one entity. The mind is wisdom, the heart is love, and the entire body as one is strength. It also describes the greatest aspect of human society: "family." The superior human family must strive for wisdom, grow to have eternal love, and conquer with strength to help unify superior humans. This philosophy was created to conjoin Judaism, Christianity, and Islam into the science of the one unified truth.

In Quantum Mahdiism Islam is the "wisdom" of life. It teaches unity and that deception is the ultimate weapon against humanity. Everything is as one. Let the superior humankind come together to grow, conquer, and strive for peace on earth for all eternity. Christianity is the true power of "love" and eternal life. Judaism is the strength that give us the will to survive. This is the grand unification. Spacetime Quantum Thermonuclear Humanity!

(Image from clipartof)

Hinduism, Buddhism, Taoism and Shintoism are all spiritual religions that are related to martial arts and meditation in nature. Hinduism is the oldest religions known in recorded history with a vast number of diverse teachings and an ancient martial art called kalarippayattu. Buddhism derives from Hindu meditation and interconnectedness with all quantum energies in nature. It is the founder of one of the most popular martial arts named kung fu. Kung fu is "the way of motion," which is "life." It has many different styles and helps tame the infinite nature of the human mind. Taoism is the "ultimate way." It practices meditation with the sole purpose of connecting to a greater source in nature. Taoism is related to a martial art named Tai Chi meaning "the ultimate fist." Those who practice Taoism does so with the intentions of connecting to the great quantum energy source. Shintoism, or Shinto-Buddhism is a superior form of meditation and martial arts that is known as "the way of the shin or gods." It is also unique in its structure and practices. It is related to Kung fu and Aikido forming a higher style of Shamploo or mixed martial art.

(Image from New Yorker)

It is time for us superior humans to project our minds beyond the things we believe in and experience on our planet. No human should be left out in this new world we live in. Like Steven Hawking said, "man must look to the stars to seek refuge, so man must seek refuge in himself." Us superior humans must join together in volition and overcome competition. Superior humans are destined to take a journey together into the deep sea of spacetime and explore the stars. We, us together as one, was always one with intelligent design. Those who seek wisdom dare to explore the nature of multiple universes based on superior human conscious. We must hold our religions together in independence and work together to continue the progression of the ancient alien superior human species. No religion is greater than the other. All humans have a psychological multiplex of aspects to think in multiple dimensions. Grow a multiple dimensional form of thinking to see pass the social dilemma of religion.

(image from civilrights-liberationcommand1)

Science

chapter: 3

"Three is for the trinity of science, technology, and engineering."
Chakotay B. Johnson

Science is a strict and theoretical and dynamical discipline that explains the observable universe and superior humans. It provides facts why they should not fight over religion and opinionated bias. These indisputable facts in science gives superior humans the chance to choose their ultimate destiny. Peace and unity, or destruction and separation. Science, in the right minds, reduces violence. The sacred knowledge in this book should never be undermined.

(Image from Wikipedia)

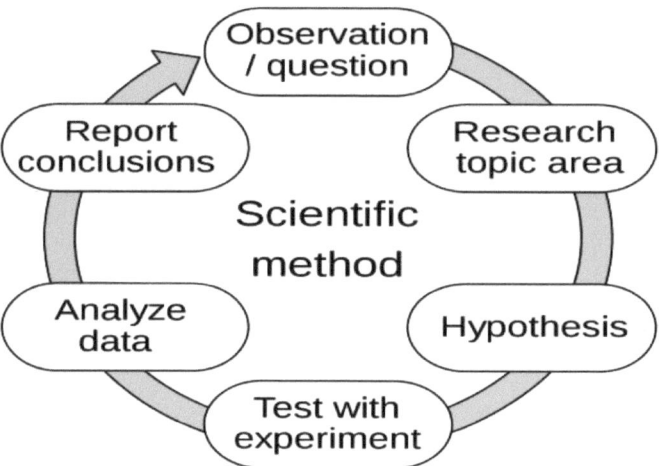

(Image from google)

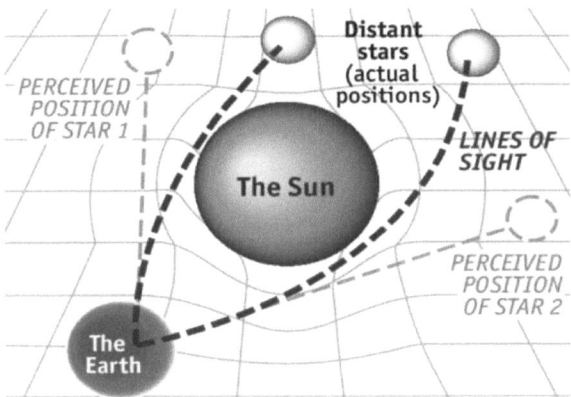

The theory of special and general relativity is the highly accurate theoretical science of space, time, matter, and electromagnetic energy. Special relativity shows that the laws of nature are the same for all frames of measurement when an object is not accelerating. No object can move faster than the speed of light, and all macrocosmic events in spacetime have an infinitesimal difference in their measurements.

$$R_{\mu v} - \frac{1}{2} R\, g_{\mu v} + \Lambda\, g_{\mu v} = \frac{8 \pi G}{c^4} T_{\mu v}$$

General relativity describes gravity as a property, of the flat spacetime field, created by the curvature of mass. The more an object accelerates the heavier it becomes. Light is bent by the curvature of flat spacetime. The progression of time slows down and the object of interest contracts due to a change in its speed. Electromagnetic energy decreases as it propagates from a gravitational system. The frequency decreases as the wavelength are elongated creating the gravitational redshift effect. Gravitational blueshift in energy is vice versa. Gravitational lensing is how light is bent around distant objects viewed by the observer. Singularities are known as black holes and are extreme disturbances in the flatness of spacetime.

Lastly, Einstein field equations states that the metric tensor equals the stress energy tensor.

(image from evolution foundation)

Organisms all throughout spacetime have been evolving forever. Terrestrial thermonuclear solar humans come from single cell organisms that arose from the rare earth chemical elements interacting with electromagnetic energy. All species in the observable universe came from conscious primordial nucleosynthesis. Astrobiologically, all organisms in the terrestrial zone of a star are subject to evolutionary changes. Genetic and Molecular biological make up of any species of organisms depends on the time spent in the surrounding environment. It can vary greatly. The multiverse is teeming with biodiversity. It is a battle of the fittest and strongest on planet earth and other planets and alternate earths where life exists. The occurrence of all life is rare and is not by accident.

Quantum consciousness of the mind does not experience time like photons. Only the physical brain does. So, like with photons, the past, present, and future are in simultaneity. This supports all religious notions of the origin of superior humankind.

(image from online-mods)

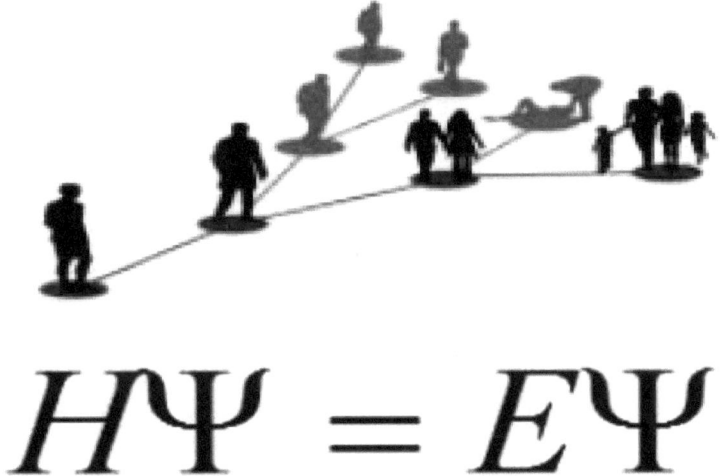

$$H\Psi = E\Psi$$

The theory of multiple universes based on the decision of quantum consciousness is the most accurate interpretation of quantum mechanics. It is called the Hugh Everett interpretation. The most basic form of this complication is the Hamiltonian-Eigen quantum oscillator equation. When an object with quantum consciousness come to a decision the universe that exists within and outside our minds split into two. So, there are almost an infinite number of universes that are created from the decisions within our imaginations. Quantum superposition is when a particle is in multiple states of position until observed. Quantum entanglement is predictable action at a distance. It is concluded that life is based on our position of decision.

(image from google)

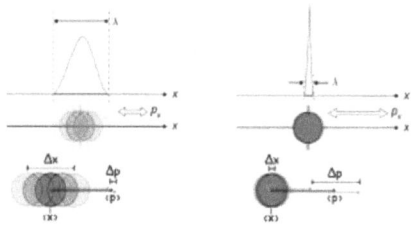

(Image from laserfocusworld)

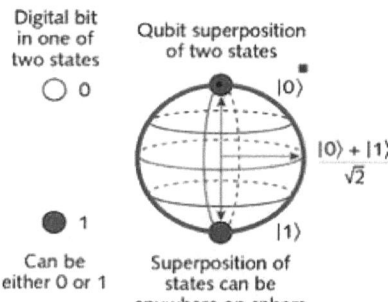

The entire universe is quantum information including all of superior humans. Everything exists as light. More advanced quantum computers of the future will be able to physically shape and simulate our reality.

(Image from civil rights library)

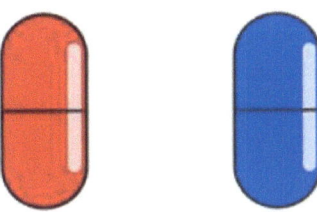

The theory of the quantum information matrix is the most bizarre and astonishing configuration of facts about our reality. The proton is a 1 and the electron is a -1 and the neutron is a zero. It is proven by the beloved physicist Marie Curie that colliding a proton and electron creates the neutron. This is what everything in physical reality is made of. Us superior humans of the new world order have a choice to work together in massive numbers to create our own reality or stay stuck in an illusion of bloodshed, violence, and suffering. The red pill is the pill of fire and motion that moves time forward. The blue pill is the pill of ice and being frozen in a time illusion. Choose wisely superior humans.

(image from civilrights library)

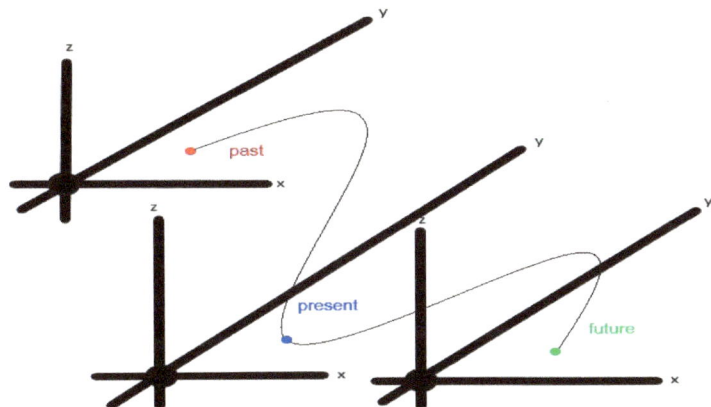

In the vacuum of spacetime ∇x, v is equal to the speed of the quantum information and c is the speed of light and π = 3.1415926535. Quantum information can travel to the past or future from the present with transdimensional linear wave propagation. Waves moving at the speed of light transmits quantum information to the future. When the speed of light is transcended by an infinitesimal fraction the barrier is broken and a tachyon is released and transmits quantum information to the past.

$$v \leq c$$

$$\nabla \frac{dx}{dt} = \frac{\pi}{v} + c$$

So,

$$c < \nabla \frac{dx}{dt}$$

(image from animescience101)

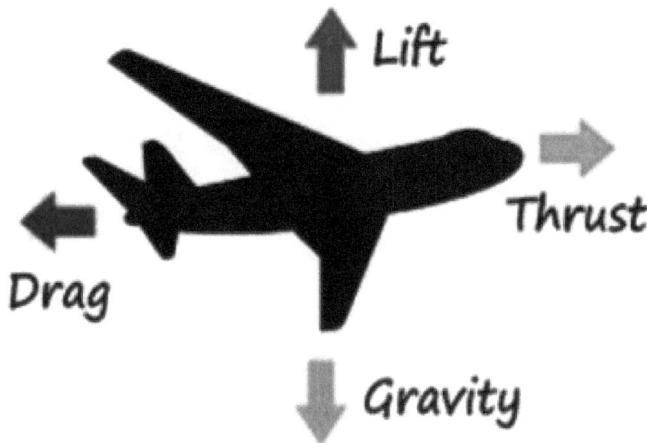

Chapter: 4
Aerodynamics

"Four is for space, time, matter, and energy." Chakotay B. Johnson

The three fundamental laws of aerodynamics are Isaac Newton's three laws of motion. Every aspect of an aerodynamical object's movement through the fluid of air and the vacuum of space are described with the three laws of motion. The aerodynamical science of flight and propulsion through the fluid of air and the vacuum of space gives rise to a more complicated system of mathematics and theory. The Bernoulli principle is one of the main advancements in aircraft technology. Lift, weight, thrust, drag, and propulsion are all a part of the theory of aerodynamics. Lift is an upward force. Weight is the force of gravity. Thrust is the forward or propulsion force. Drag is the force working against thrust. Bernoulli's principle is fluid-pressure increase and decrease.

(image from civilrights library)

$$v = \{0 \text{ or } C\}$$

$$m\frac{dv}{dt} = 0$$

v = speed in a certain direction

m = mass

t = time

C = constant speed in a certain direction

The first law of aerodynamics is inertia. An object at rest or at a constant velocity tends to stay at rest or in motion unless acted on by an external force. An example is a ball sitting on the ground will stay at rest until it is kicked by someone. Another one is the earth will stay in orbit unless its speed rapidly increases at a point tangential to its orbit.

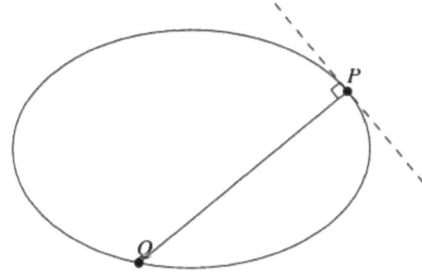

(image from civilrights library)

$$F = m\frac{dv}{dt}$$

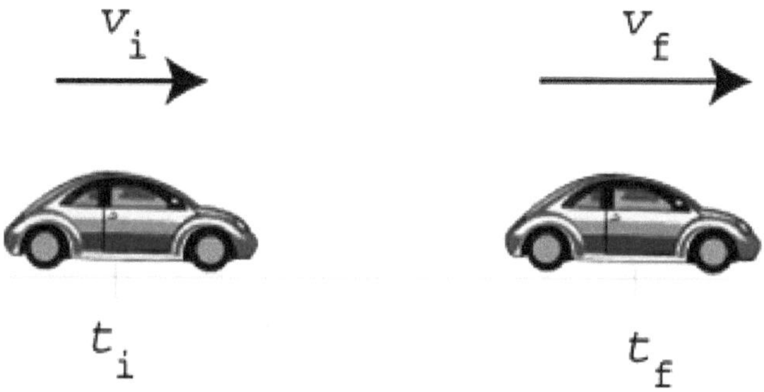

v_i →

t_i

v_f →

t_f

F = force

m = mass

v = speed in a certain direction

The second law of aerodynamics is force of acceleration. The mass of an object multiplied by the rapid change in velocity with relation to time is equivalent to force of acceleration. An example is a car moving down the road with applied gas to its engine in relation to the time the car's velocity starts to increase equals the car's force of acceleration.

(image from civilrights library)

$$F = -F$$

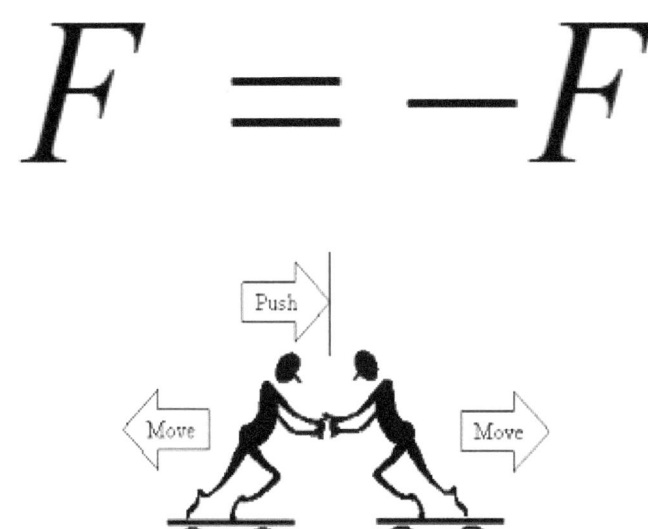

The third law of aerodynamics is the action of force creates an equal and opposite reaction force. An example is two skate boarders pushing toward each other will stop until one of their push forces becomes greater. Another example is an airplane that is pulled forward with its propellers experience opposite forces.

(image from aerospace)

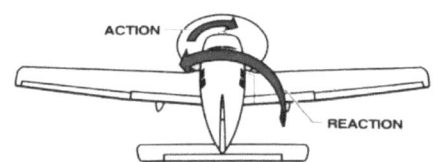

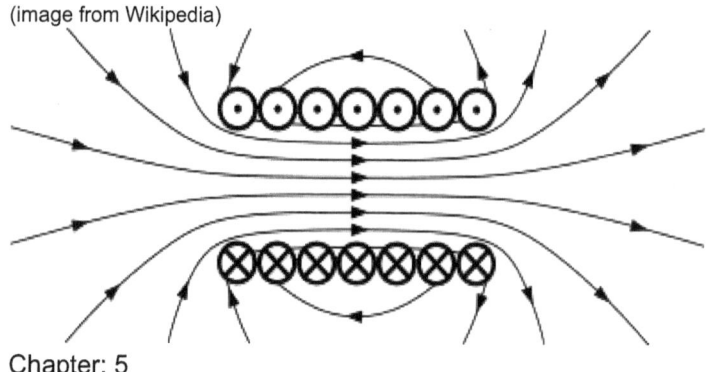
(image from Wikipedia)

Chapter: 5
Electrodynamics

"Five is for women as wise queens." Chakotay B. Johnson

The four fundamental laws of electrodynamics describe all phenomena in our universe. A scientist named James Clerk Maxwell discovered the simplicity of a more complicated system of these laws. They are therefore called Maxwell's equations. Everything in our observable universe is electromagnetic in its composition. The brain is the most advanced electrical object in existence. It cannot be recreated, instead it could be simulated and emulated and resurrected. These equations have blessed our societies with a trove of electrical and magnetic technologies.

(image from civilrights library)

$$\nabla \cdot \mathbf{E} = \frac{\rho}{\varepsilon_0}$$

∇ = gradient operator

$\mathbf{E}$ = Electric field ρ =

charge density ε_0 =

permittivity

The first law of electrodynamics shows how an electron creates a uniform field of force pulling inward. Another approaching electron will be repelled if the electron is excited or at a higher level of energy than it, and vice versa.

(image from civilrights library)

$$\nabla \cdot \mathbf{B} = 0$$

∇ = gradient operator

B = magnetic field

The second law of electrodynamics shows how a magnetic field is produced by two poles of opposite charges and circulates. The force of magnetism is zero without an opposite pole. A good example is a one-way magnet does not produce a magnetic field of circulating force.

(image from civilrights library)

$$\nabla \times \mathbf{E} = -\frac{\partial \mathbf{B}}{\partial t}$$

∇ = gradient operator

E = Electric field B =

magnetic field t =

time

∂ = partial change

The third law of electrodynamics shows how a changing magnetic field creates a circular electric field. It also shows the undulating dualistic propagation of a concentration of electrons and a force field of magnetic energy. Electromagnetic energy that exists as light and moves in a quantum of synchronized periodic motion. Modern theories of electron motion call this phenomenon particle-field duality. The electron behaves like an object and force field.

(image from civilrights library)

$$\nabla \times \mathbf{B} = \mu_0 \mathbf{j} + \frac{1}{c^2} \frac{\partial \mathbf{E}}{\partial t}$$

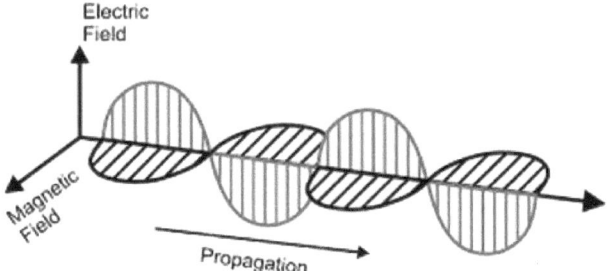

∇ = gradient operator

c = speed of light energy

E = Electric field

t = time

B = magnetic field

μ_0 = permeability

∂ = partial change

j = current density

The fourth law of electrodynamics is the reverse law of the third law. It shows how a magnetic field interacts with a flow of electrons to create electromagnetic waves. The magnetic field is perpendicular to the electric field and moves through the vacuum of space with a speed of approximately 299,792,458 meters per second. Electromagnetic waves propagate through space as time progress. This theoretical science revolutionized our planet with technologies like transformers, electrical generators, inductors, electrical engines, computers, phones, and electrical turbines.

(image from walpaperaccess)

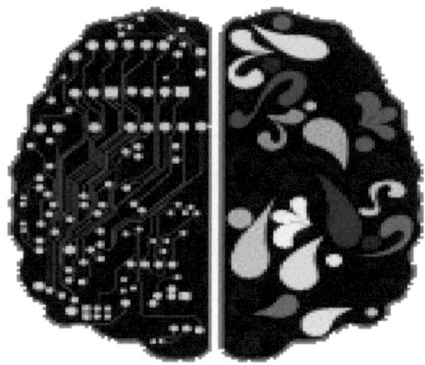

Chapter: 6

METADYNAMICS

"Six is for men as strong kings." Chakotay B. Johnson

The three fundamental laws of pseudo-quantum metadynamics describes the informational multiverse of the mind. In this theoretical science, the mind is seen as a substance of light energy in relation to the functionality of the physical brain. The functionality of the brain is the most complicated and puzzling system to study in the history of science. It is not fully accepted as a quantum mechanical system. Research and testing are currently in progress to discover if it is. The brain functions as a network of interconnected neurons called brain cells. Neurons store human memory. Memory is what give us a perception of reality in spacetime's unified continuum. In this theory, memory, whether short-term, or long-term, is the main property of the mind or psyche. It gives superior humans the ability to function, grow in ideas, and survive. Information processing is the memory. Memory can be lost due to brain damage. The interconnected neural network of the brain and the rest of the body's nervous system communicate at almost instantaneous speeds. The superior human mind, depending on the complexity of its intelligence, concentration, and challenge of learning, radiates an explosive multitude of thoughts and brain activity. Excluding signals passed around the nervous system to perform external tasks in the environment, the mind is in a multiplicative state of simultaneity. Picture you and me playing hockey. The probability of that actually happening as you read the words of this book's speech, is happening within your mind's informational multiverse sea.

(image from civilrights library)

$$M = M_i$$

M = 0kg

M_i = 1sub

The first law of metadynamics shows how the mind exists as a primordial light energy that can think with zero-mass and is equal to one substance. It is nicknamed the M & M law. The mind's information exists as one substance with the continuum of spacetime. Unconsciousness is the foundation of the mind. Consciousness materializes and physically shape multiple physical realities based on decision and imagination. The two relics exhibit the property of quantum entanglement. For example, a human can be definite and make unconscious decisions.

(image from civilrights library)

$$\frac{dM_i}{dt} = e^{T(L,C)}$$

dM_i = constant flow of information as one(1sub) substance equals 0kg

dt = progression of time   T = thoughts

L = logical thoughts   C = creative thoughts

$e^{T(L,C)}$ = 1sub

The second law of metadynamics shows how the constant flow of the informational mind as time progress has zero-mass and is equal to a natural exponential function of logic and creative thoughts. This equation shows the relationship between the mind's mass and substance. All the information in the mind is unified into one simultaneous function that has the quantum behavior of light. 0kg = 1sub. The integral-derivative transformational relationship confirms the first law of metadynamics.

(image from civilrights library)

$$\frac{dM_i}{dt} = M \qquad \int_a^b M\, dt = M_i$$

(image from civilrights library)

$$\sum_{i=0}^{\infty} M_i = (T \cdot B \cdot E)_i$$

M_i = informational multiverse of the mind

T = thoughts

B = behaviors

E = emotions

The third law of metadynamics shows how the sum of pseudo-quantum states of the mind's stored information controls the cyclic cognitive chain of thoughts, behavior, and emotions. So, what we think controls what we do and how we feel. The pseudo-quantum superpositional multiplicative sum shows how thoughts control emotions and behavior. The three aspects exist in multiple observable states of unity. Applications of this theoretical science are transcendental meditation, astral projection, martial arts, and quantum psychology practices. To rise into a mental state of superiority and higher brain frequency, superior humans must continue learning.

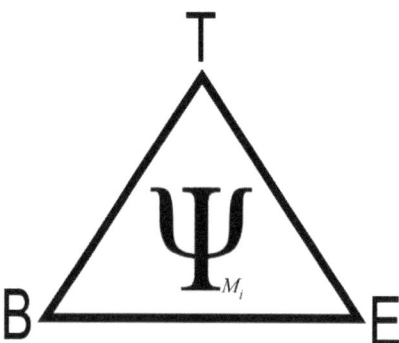

(image from civilrights library)

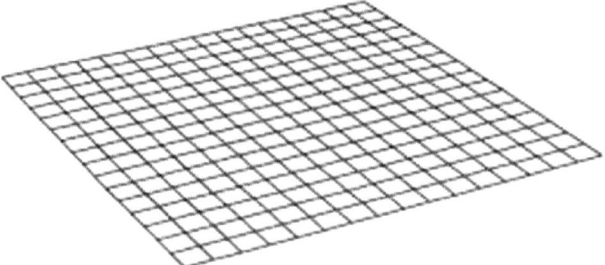

Chapter: 7

Timespace Dynamics

"Seven is for luck and blessings." Chakotay B. Johnson

Timespace, spacetime dynamics is the science that multiple theories and experiments are built upon. It shows that spacetime is a flat vacuum and unified continuum on extremely large scales. Objects with mass distorts this flat vacuum into curvatures to transverse gargantuan distances in shorter periods of time. The faster an object move through space, the more time slow down, and its length contract. Scientists can build rods, disks, and warp-drives of high propulsion with this technology. We are destined to travel the spacetime.

(image from civilrights library)

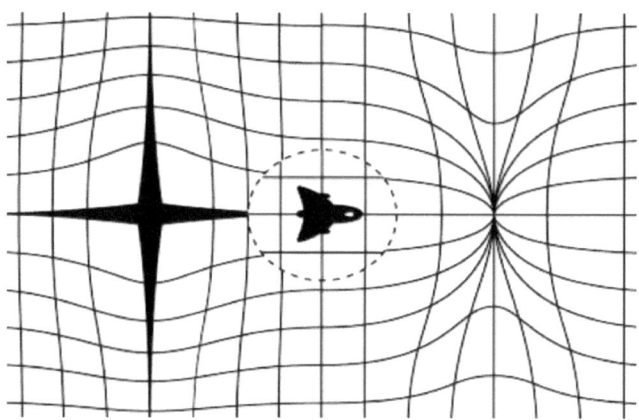

(image from civilrights library)

$$m(x,y,z,t) = 0$$

m = mass x =

length y = width

z = height/depth

t = time

The first law of timespace dynamics shows how the origin of any system is the heaviest thermodynamical body of mass which creates length, width, height/depth, and the progression of time.

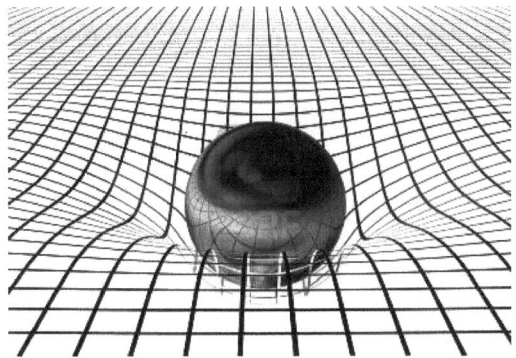

(image from civilrights library)

X = disposition

v = speed in a direction

t = time

The second law of timespace dynamics shows how the disposition of a thermodynamical body of mass equals the velocity multiplied by time. Velocity is constant. Velocity cannot be equal to zero due to the third law of timespace. So, this book exists today because we have been in motion since the beginning of time.

(image from civilrights library)

$$\acute{\phi} = 0 \qquad \phi = \frac{dt}{\sqrt{\dfrac{X_1 X_2}{v^2}}}$$

φ = flat spacetime

dt = difference in time

X_1 = initial disposition

X_2 = final disposition

v^2 = symmetrical velocity

i = square root of negative one (imaginary number)

The third law of Timespace dynamics shows how everything comes from the sun. Behold, the law of everything in our observable reality. Everything in our macroscopic reality can be described with the approximated ratio of this equation. Quantum Thermonuclear Timespace dynamics unifies all existence into one creation. The unified spacetime equation holds the following facts. Timespace is curved by mass. Timespace has an origin in every star system. Timespace, over vast distances, is flat, and ultimately an illusion. Timespace will expand and progress forever. Our greatest imagination is a real physical reality.

(civil rights library)

This is a picture of spacetime over a vast distance of hundreds of millions of lightyears showing distant galaxies and stars.

(image from civilrights library)

step 1. $\frac{X}{v} = t$

step 2. $\frac{t}{\frac{X}{v}} = 1$

step 3. $\frac{t^2}{\left(\frac{X}{v}\right)^2} = 1^2$

step 4. $\sqrt{\left(\frac{t}{\frac{X}{v}}\right)^2} = \sqrt{1^2}$

step 5. $\frac{d}{dt}\sqrt{\frac{t^2}{\frac{X_1 X_2}{-v^2}}} = \frac{d}{dt}(1)$

step. 6 $dt = \frac{X_1 - X_2}{v}$ $\phi = 0$ $\phi = \frac{dt}{\sqrt{\frac{X_1 X_2}{v^2}}}$

(image from Caltech)

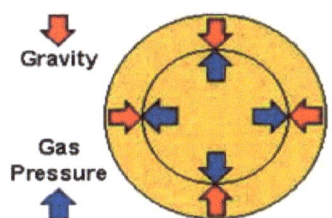

The above steps are how the unified spacetime equation was discovered. It was derived from the disposition-velocity-time equation using Einstein's square derivation process after it was put into a uniform ratio. To accurately calculate something, you have to square it, then take the root of its squared form. The above equation is the unification of spacetime and mass-energy. Everything in our observable existence is one entity. It conforms with vector motion cancelation, and gas-pressure-gravitational balance in the sun, and other dynamical models.

(image from civilrights library)

Chapter: 8
Thermodynamics

"Eight is for the four quadrants of infinity." Chakotay B. Johnson

Quantum thermodynamics is the foundation of our reality. It is indisputable facts about the properties of heat and entropy or the measure of disorder in all observable objects. Thermodynamics created quantum mechanics and quantum field theory. Things like the uncertainty principle and the correspondence principle was discovered during research of this dynamical science. Quantum field theory is the most recent and interesting of these sciences. It merges quantum mechanics, classical mechanics, and special relativity into advance models of the behavior of subatomic particles that exist as energy fields. Quantum uncertainty and correspondence principle are fundamental to the understanding of our reality. This science describes us as ancient alien quantum thermonuclear humans.

(image from civilrights library)

$$dU = Q - W$$

dU = change in internal energy

Q = quantum heat energy

W = work

The first law of thermodynamics shows how internal energy of any object equals a quantum of light added to the system, minus work done by the system. Energy cannot be created or destroyed, but only conserved and transformed.

(image from civilrights library)

$$dS = \frac{dQ}{T}$$

dS = measure of disorder of two states in a system

dQ = quantum heat energy transfer

T = temperature of environment

The second law of thermodynamics shows how the change in the entropy or disorder of a system equals the ratio of its transfer of heat energy to its surrounding temperature. Any reversible and irreversible quantum heat process can be reversed.

(image from civilrights library)

$$Q = mc\,dT$$

Q = quantum heat energy m

= mass in kg

c = quantum heat capacity

dT = change in temperature

The third law of thermodynamics shows how the entropy of a system approaches a constant value as the temperature approaches absolute zero. The measure of disorder in a system is typically zero if the temperature is absolute zero depending on the number of ground states. Everything is electromagnetic quantum heat energy. Exotic forms of matter with unconventional properties does exist.

(image from astrofamily)

Chapter: 9

What is the fate of humanity?

"Nine is for the superior life inside the womb of a woman."
Chakotay B. Johnson

This is a world of the new order. The scientific order. Thermonuclear solar humans must rise into a state of superiority and reshape the course of history. Superior humans must return to the facts of the light. We must join the adventure of learning to build infinite information into the immaterial pseudo-quantum substance of the mind. It is up to us to build renewable energy and safe our home planet from pollution, disaster, and its enemies. The more superior human minds working together is for the better. Let us sway the motion of our attention towards reading and education with pure dedication and determination. We are really one great new world nation. This book is the grand unification. Steven Hawking said, "because there is a law such as gravity, the universe can and will create itself from nothing." The unconscious nothingness of the mind will strive to become conscious once again. Build you a team of minds to explore the sciences so we can rise into a state of total awareness.

Separation will make time slowly pass us by. Working together in unity will push time forward at the speed of light. Superior human consciousness is what created everything in our reality.

Ultimately, there is no god, superior humans must harness the power of unity and peace. Us together is god who is king. This book is the information that is stored within us. United we stand, divided we fall. We are the bright and shining stars. New World Order: spacetime quantum thermonuclear humanity!

About the Author

(image from civilrights library)

Chakotay

Bobbie James Johnson is a man with the soul of a child and the heart of a king. He is a serious man about learning, peace, and unity. He grew up on the south side of Chicago where he studied the physics of our reality. He is now a theoretical and dynamical physics and computer science student in the Minnesota state college system. He dreams of creating the best video games and protecting the civil rights of minorities. His greatest dream and imagination are to meet his deceased cousin Andrew Marcus Powell again and they play video games forever.

www.ingramcontent.com/pod-product-compliance
Lightning Source LLC
Chambersburg PA
CBHW040244220526
45473CB00001B/358